A SMALL FABLE OF PERSONALITY'S *big* FIVE

JIM EXLEY, PhD, PATRICK DOYLE, PhD
& W. KEITH CAMPBELL, PhD

Copyright © Professor OCEAN LLC 2021

Professor OCEAN LLC. All rights reserved.
Printed in China.
No part of this book may be used or reproduced
in any manner whatsoever without written permission.

For more information:
ProfessorOCEAN@Professor-OCEAN.com

731 Duval Station Road #107 #283
Jacksonville,
FL 32218

ISBN 978-0-578-30040-5

PROFESSOR OCEAN™
A SMALL FABLE OF PERSONALITY'S *big* FIVE

JIM EXLEY, PhD, PATRICK DOYLE, PhD & W. KEITH CAMPBELL, PhD

INTRODUCTION

Welcome to the world of Professor OCEAN! The purpose of this fable is to introduce you, the reader, to the construct and science of what's commonly referred to as the "Big Five" in personality traits. Or, as we like to call it, OCEAN.

OCEAN is an acronym that represents the five domains of personality:
- Openness
- Conscientiousness
- Extraversion
- Agreeableness
- Neuroticism

In the tradition of other great business fables, the goal of Professor OCEAN is to convey a massive amount of science and wisdom in a short amount of time—possibly just one sitting! Understanding that everyone may explore to different depths of this rabbit hole, we've written this fable in layers. Some of you who've dabbled in the field of psychology might recognize some classic research findings and familiar names. Those of you steeped in research may notice that each lesson is referencing well-researched topics and key meta-analyses. And others of you may simply enjoy reading the fable and meeting a new friend and guide in Professor OCEAN.

Regardless of how you approach the subject of personality science and this book in your hands, there is no need to have a PhD in psychology to see and apply the power of what you'll learn here. Our hope is that this little book will help us all understand and appreciate each other a little more. That it will help us discover more about who we are and how we're wired. That it will create a language that's accessible to us all.

Enjoy!

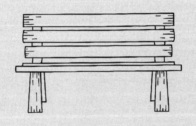

CHAPTER
One

PERSONALITY MATTERS

Two old friends, William and James, met at a sidewalk coffee shop on the morning after their school reunion. James was already seated when William arrived a few minutes late.

"Good morning, bud!" James greeted his friend. "Great seeing you last night."

"I'm not so sure it's a great morning," William replied.

He was clearly recovering from a late night, and quickly placed a much-needed coffee order when the waiter arrived. Then, he turned his attention back to James.

"I'm glad we've got some time to catch up," James said. "Last night's mixer was great but not so much for meaningful conversations. So, what's new with you? You still dating Stacey?"

William shook his head. "Nope. That ended

a few months back."

"What happened?" James asked, clearly surprised. "I thought she was the one!"

"So did I," William admitted. "It was same ol' thing. She said I was lots of fun, but a little too all over the place for her. She wanted me to calm it down. Plus, the arguing was pretty intense. I just decided it was time to move on."

James was shocked. Stacey's a catch, and he thought William knew that. Why would he ever think it was "time to move on" from a girl like that? Seeming to sense his friend's disapproval, William moved the conversation quickly to a new subject.

"So, what's new with you and the family?" he asked.

"All good on my end. Caryl is in the middle of writing another book. Henry is playing travel baseball most every weekend, and Kristen is heavily involved in school theater. We spend lots of time chasing the kids around, but it's fun!"

Fun? To William, the life James described sounded exhausting!

"My company is doing great," James

continued. "We're up to 10 employees and just signed a new office lease right around the corner. I feel like we're doing great work, helping folks, and enjoying the ride. Plus, it allows me to spend lots of time with Caryl and the kids, which I love!"

William laughed. Years ago, James left the corporate world to start his own firm. Since then, he'd made a decision to keep his company small in order to preserve time with his family. Truth be told, it was a decision William never understood.

"Has anyone ever told you that you're wasting your talent?" William exclaimed. "It would be amazing to see what you could accomplish if you just decided to work a little harder!"

Grinning, James shifted the subject. "Yep, I might've heard something like that before. How about your job? How are things in your work world?"

"I just won salesman of the year!" William replied proudly. "I sold twice as much as the number two guy."

"And your relationship with your boss, Mariah?" James asked.

Here, William paused. "Not great. I got another 'doesn't play well with others' on my last review. But whatever. My sales numbers speak for themselves. It's her problem, not mine."

James decided to tread lightly as he made his next point.

"Arguing with Stacey, stirring things up at work…? Listen man, it sounds like that personality of yours keeps getting in the way."

"Look here, Mr. Goody Two Shoes, college is over," William said, exasperated with his friend's take on the situation. "I didn't come down here this morning to get lectured by you. Have you been talking to my parents or something?"

The question was fair. James had spent many nights in college talking to William's parents. Together, they were always trying to keep him on the straight and narrow. Feeling himself heading down the road to an argument, William pulled back. He didn't want to be an idiot to a friend who had always helped him out.

"Hey man, sorry to be so snippy," William apologized. "You don't deserve that."

"Did I just hear William say he's sorry?" James joked. "Wow! Wonders never cease!"

"I'm just sick of people trying to change me," William admitted. "I mean, of course someday I'd like to settle down. And I can't keep pissing people off at work. But where do I even start? At my age, with my baggage, change seems like more work than it's worth."

"Sounds like you need a little help learning to play well with others. I think a visit with Professor OCEAN might help," James suggested.

The mention of their old professor perked William up. He hadn't thought about Professor OCEAN in years. He honestly couldn't believe the guy was still alive; he had to be over 90 by now!

"I thought he retired and moved to the beach to surf," William said.

"Well, he did," James confirmed, "but the university brought him back by popular demand."

James paused, giving his friend the chance to consider the opportunity in front of him.

"Listen, he's actually expecting you right now," James continued. "When I saw you in action at the mixer, I had an idea you might need him. Then, serendipity took over! When I got

Personality Matters

here early today, Professor OCEAN happened to be sitting right there with his dog, Sigmund. Long story short, he said he'd be around if you wanted to swing by."

William thought about it. He had to admit, the idea of seeing their wise, old teacher was intriguing.

"Listen, I've got to run," James said, standing to leave. "But I really think you should go. Of course, I understand if you don't have the courage to head over…"

William laughed. "You had to bring my manhood into this, didn't you?"

"I'm your best friend," James yelled across the sidewalk, "I know what works with you!"

Left alone, William considered his options. This whole thing sure felt like a set up. It was almost as if James had planned it. Sure, Professor OCEAN was wise, but William hadn't always been his biggest fan. After all, the guy did give him a D in college for "low attendance." But grade aside, he also knew the Professor was extremely successful and wise, and William could use a little wisdom. His job, things with his boss, his relationship with Stacey—none of them were going as planned. And as James so boldly stated, William seemed to be the common denominator. Maybe a quick visit with Professor OCEAN could help. At the very least, it would be nice to just say hello.

Reluctantly, William made the decision. He'd stop by and see Professor OCEAN. What did he have to lose?

Personality Matters

Personality is your stable and unique combination of characteristics, experiences, and stories that influence the way you see the world. In other words, it's what makes you, you! And that's why personality matters!

William's brash and aggressive personality might help him succeed at his job, but it's also led to some problems in his life. James' more stable, agreeable personality has helped with his relationships, but he's not selling himself like William at work anymore. The point? We all have different personalities that shape our lives in important ways.

Maybe you've experienced the same thing in your own life. Like William, it might even leave you wondering if you're the problem. As you read, consider situations where your personality has worked for you and where it's worked against you.

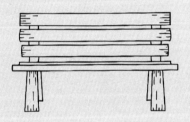

CHAPTER Two

WE CAN KNOW OUR PERSONALITY

As William got off the elevator on the 11th floor of Professor OCEAN's building, he could hear the familiar notes of Led Zeppelin's "The OCEAN" playing down the hall. It was booming rather loudly for a Saturday morning, and as he approached the door of apartment 1117, William realized it was coming from behind the Professor's door.

"There is no way this can be right," William muttered to himself. "There is no way this old dude is listening to Zeppelin."

Shrugging his shoulders and bracing himself for what he'd find behind the door, William lifted his hand to knock.

"Here goes nothing."

His knocks were met with the bark of Sigmund from behind the door. After a moment,

We Can Know Our Personality

William knocked a little louder and finally heard Professor OCEAN's familiar voice.

"Hold up! You don't have to break the door down. I'm coming. You interrupted my favorite part!"

The door swung open, and just like that, Professor OCEAN himself was standing in the doorway.

"Hello William! James said you might swing by. If you could muster the courage, of course," the Professor said with a sly grin.

William was awestruck. In some ways Professor OCEAN looked *really* old. But in other ways, William was surprised at how young he still seemed. The Professor was tall, lean, and muscular in a wiry kind of way. His face was sun-kissed and wrinkled, and he still had a full head of silver hair that matched his beard. But really, his piercing blue eyes were what made him look young. Somewhere in there, his eyes revealed a playfulness—a type of childhood excitement. He was dressed in up-to-date athleisure, including low top basketball sneakers and an updated version of the same alligator belt he had worn when William was an undergrad. He still wore the same vintage Rolex submariner on a matching alligator strap, too. A quick glance at Sigmund, his rust-colored golden doodle, revealed an alligator collar matching Professor OCEAN's look. Obviously, the old dude loved his alligator.

"How can I help you?" Professor OCEAN asked, inviting William inside.

"I'm not sure…" William trailed off.

"Well, if you aren't sure what you need, I'll get back to my day. Sigmund and I have big plans! Isn't that right, Sigmund?" He turned to smile at the dog by his side. "Nice to see you again, William."

Just like that, Professor OCEAN was moving to close the door. If he didn't do something fast, William would be left alone in the hallway.

"Wait!" William cried, stopping the door from closing completely. "I think you might be able to help me."

Professor OCEAN paused, staring at William intently. "Listen, William, a person needs to be clear on three things: who you are, what you want, and what your plan is for getting it. And by the look on your face and the hesitancy in your voice, it seems you don't know any of the three. I'm not sure I can help you."

"I know I want my life to be better," William pleaded. He was surprised by his sudden honesty, both with the Professor and with himself.

Sensing William's sincerity, Professor OCEAN nodded. "Okay, we'll start with that. Meet me on Thursday at 10:00 a.m. on the park bench in the quad. I'll have a black coffee with a shot of espresso—same as you. That's where we'll begin!"

William wasn't sure what to do. He had to work on Thursday, and he lived two hours away from the university. The meeting with Professor OCEAN was impossible for him to make. But when he told the Professor as much, he was met with only a laugh.

"I'm sure Mariah will let you take a vacation day," Professor OCEAN said knowingly. "I'll see you on Thursday."

And with that, Professor OCEAN closed the door. Within seconds, the music picked up right where it left off. Dumbfounded, William stood in the hallway for a few moments before heading to his car. On the drive home, he wondered if he was living in the twilight zone. What did Professor OCEAN remember about him from undergrad? How did he know what his coffee order was? And how did Professor OCEAN know his boss, Mariah?

William was reluctant, but ultimately, his resolve remained: He was going to take the journey with Professor OCEAN.

We Can Know Our Personality

Understanding your personality requires honesty. William had been telling himself the story that his life was going pretty well. But in his meeting with the Professor, he realized that wasn't the whole truth. Maybe his personality had created some problems.

Until William takes a closer look at who he is, he won't be able to live the life he wants to live or become the person he could become.

We Can Know Our Personality

The same is true for you! Before becoming the person you want to be, you must figure out the person you are. And that demands an honest look at your life.

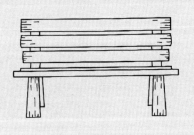

CHAPTER Three

OCEAN PERSONALITY

When William arrived at work (15 minutes late as usual), there was a note from Mariah asking him to stop by her office. So, he walked that way, none too pleased to be going. As the top salesman in the company, he didn't understand why she'd breathe down his neck for being 15 minutes late. He was doing his job, wasn't he?

When he arrived, he found Mariah's door open, and interestingly, she greeted him with a smile on her face.

"I heard you met Professor OCEAN," Mariah said.

William was shocked. "Who told you that?"

"He did, of course. Are you going to take him up on his offer to meet this Thursday?"

It seemed as if everyone in William's life was

running surveillance on him.

"What else did he tell you?" William asked.

Mariah laughed. "That's about it. Well, other than the fact that your eyes were bloodshot and your face looked like you'd seen a ghost."

"I was just there for a minute," William said. "How could he pick up on all that?"

"Come on William, he's one of the top psychologists in the world," Mariah replied. "So, are you going to do it? Are you going to meet with Professor OCEAN?"

William wasn't sure the meeting was worth one of his precious vacation days. And when he told Mariah as much, she grinned in reply.

"William, hear me clearly. You need to do this. Remember the 'doesn't play well with others' comment from your review? You are a top performer, but if you don't make some changes, I'm not sure how much longer I can keep you. Corporate is breathing down my neck about your behavior. The word is that you are a natural born leader, but you are leading in the wrong direction. I think Professor OCEAN can help."

Hearing this, William knew his next step.

Even if it only got the company off his back, a meeting with Professor OCEAN would be worth it.

"All right, I'll go see him, at least this once," William told his boss, turning to walk away.

"Good decision," Mariah replied. "And William? Don't be late."

After making the two-hour drive for their first meeting that Thursday, William was in a less than pleasant mood. He grabbed the two coffees just as Professor OCEAN requested and walked across the street past the statue of the school mascot. As he rounded the corner (five minutes early, for the record), there sat Professor OCEAN, legs crossed reading the newspaper with Sigmund by his side.

Professor OCEAN greeted him with a twinkle in his eye. "Well, you decided to show! And here I thought I might have to buy my own coffee today."

As William sat down, the Professor got right to the point. "So, how can I help you?"

"Well, Mariah says I don't play well with others," William began reluctantly.

"And that's what caused you to drive two

hours to have coffee with an old man?" Professor OCEAN nudged.

"Well, I'd like to keep my job," William continued. "And it's not just my job, I guess. The girl I was dating also mentioned I need to grow up. She told me to give her a call if I ever wanted to take life more seriously and stop arguing over meaningless stuff."

Professor OCEAN nodded, taking a pause that was almost awkwardly long from William's point of view. When he finally did speak, he seemed to change the subject.

"Did I ever tell you how I got my name?" Professor OCEAN asked.

"I thought your parents gave it to you," William quipped.

"No, it's an acronym."

"An acronym?" William asked, feeling slightly stupid. "For what?"

"Do you remember my lesson on 'Big Five' personality?" Professor OCEAN asked. "NEO-AC?"

"It sounds familiar," William admitted. "I always thought you were talking about the *Matrix* dude—Keanu Reeves. I'm not sure if you remember, but I got a D in your class, Professor. It seems I didn't learn that much."

Professor OCEAN smiled. "No, William, you earned that D."

William shrugged. "I guess I had that coming."

"Don't worry, lots of students had trouble remembering NEO-AC. That's why I rearranged the letters to spell OCEAN. When my books began catching on, somebody labeled me Professor OCEAN, and it stuck."

William was almost too embarrassed to ask. "But what does it stand for?"

"The five traits of personality, of course," Professor OCEAN replied. "Openness, conscientiousness, extraversion, agreeableness, and neuroticism. Together, they're probably the most meaningful and useful contributions of personality science in the last 50 years."

From there, Professor OCEAN broke down the meaning of his name—the acronym that helps his students understand personality.

"The O represents openness, the trait that tells us whether a person likes new ideas or prefers the status quo," the Professor began.

"Most traits are measured on a scale of high to low. In this case, low means a person is very traditional in their thinking and not open at all, while high means they're wide open."

"But higher is better, right?" William interjected.

"Not necessarily. Personality can create challenges at both the high and low ends. I knew some people back in Berkeley who were too open for their own good!"

Already learning, William was eager for the Professor to keep going. "And the C?"

"That's conscientiousness," the Professor said. "Like, if a person makes their bed or not."

At this, William chuckled. Surely being conscientious wasn't just about making the bed in the morning! When he said as much to the Professor, he was met with a laugh.

"My job is to get people to remember things, William. And since you already sat through some of my lectures in undergrad and none of it appears to have stuck in your brain, I figured I'd put the cookies on the lower shelf for you this time. So, yes, for you, at this point in your life, conscientiousness is about making the bed.

Making your bed takes dutifulness, orderliness, industriousness, and achievement-striving."

William conceded, knowing he'd been put in his place.

"Got it," he said quietly. "So, what about E?"

"E is for extraversion, meaning if a person is outgoing and gregarious or quiet and reserved," the Professor explained. "This is the trait most people are familiar with, but extraversion isn't only about being a social superstar. It also includes drive and energy level, so leaders and builders are generally high in E. A person lower in E will be more comfortable on the periphery of social settings and less willing to draw attention to themselves. In a word, they're quiet."

This one made sense to William, so he leaned back in, ready for more.

"And A?" he asked eagerly.

"Agreeableness. So, is a person argumentative or easy to get along with?"

Without question, William knew where he was on that scale.

"A person high in A is going to seem nice," the Professor continued. "They may not speak up even when they disagree, which could actually cause some problems. A person low in A could be called argumentative and may disagree in conversations simply for the fun of it. Sometimes this is misinterpreted as anger, but in fact, it's just their personality."

At this point, William was interested. He'd experienced this struggle so often with the people in his life, primarily Mariah and Stacey. At times, they'd both thought he was mad while he only believed he was challenging their points of view. Now, it made more sense.

"And the N?" William asked, wanting to know more.

"N stands for neuroticism."

Before the Professor could continue, William cut in. "That doesn't sound fun."

"Well, another way of describing neuroticism is its opposite, which is emotional stability," Professor OCEAN explained. "It's the only item that is reversed, because we already have an E on our list."

William wondered if this was why his mom would get so fired up about him tracking mud into the house. To him, it wasn't a big deal. But to her, it was huge!

When he shared this thought with Professor OCEAN, he got confirmation. "Yes, being high-strung is one version of neuroticism that we often think of, but it can also be feeling blue. We usually refer to it as the tendency for someone to experience negative emotion."

Again, William nodded. He was really starting to see it.

"So, what does high and low N look like?" he asked.

"Neurotic sounds like a negative word, but that isn't necessarily true," the Professor began. "Like all traits, extreme highs and lows can cause problems. With high neuroticism, we are more likely to find psychological disorders like anxiety or depression. But neuroticism in mild doses serves as a warning device. I still get a little anxious the night before a big talk, which motivates me to double check my materials. I've averted a couple of disasters that way."

He paused for a moment, watching the information sink in for William.

"I think of the OCEAN traits like dials. They can go from a very high setting to a very low setting, but most of the time, people are going to be somewhere around the middle. In our culture, the ideal setting is a little bit open, a little bit conscientious, a little extraverted, a little agreeable, and a little emotionally stable (or a little low in neuroticism). But it's really more complicated than that. Sometimes, we need more of a specific trait. For example, when we have kids, we are pulled to become more conscientious and agreeable. When we go on the job market, we want more extraversion. Different contexts call for different personality traits. It's not always as simple as it seems."

William realized Professor OCEAN was right. This whole thing wasn't that simple at all. How was he ever going to grasp it?

Almost as if reading his mind, Professor OCEAN gave an answer. "To help you, William, I'm going to give you some homework. You're going to do a scavenger hunt for personalities!"

"And how do you propose I do that?" William asked, surprised.

"Next week, I want you to come back with an example of each OCEAN trait," Professor OCEAN explained. "Think of it like bird watching, but for personalities. A mental snapshot of each will work just fine."

Before William could ask any questions, Professor OCEAN took his last sip of coffee and stood to leave. "Come on, Siggy! We're going to be late for our next meeting!"

William looked at his watch. He had only been there 37 minutes.

"Wait, that's it?" William exclaimed. "I spent two hours driving for only 37 minutes?"

"You get what you pay for, William," the Professor shouted as he walked away. "See you next week—same time, same place. Oh, and make mine a large coffee!"

And with that, Professor OCEAN was gone, Sigmund trotting happily by his side.

OCEAN is a great way to understand personality. Professor OCEAN told William about the "Big Five" traits: openness, conscientiousness, extraversion, agreeableness, and neuroticism. As a reminder, here's a brief description of each.

Openness: Openness to experience has elements of fantasy and creativity and an interest in beauty and art, as well as more intellectual interests, like philosophy and the world of ideas. Very high open people can be described as creative or unusual, while low open people can be described as conventional or traditional.

Conscientiousness: Conscientiousness has elements of dutifulness, organization, industriousness, and achievement-striving. The association of conscientiousness with discipline—showing up on time and working hard—is why many employers look for it. People lower on the conscientiousness scale may be described as disorganized and spontaneous. Too little conscientiousness may cause a person to appear lazy, disheveled, and even undependable.

Extraversion: Extraversion has elements of enthusiasm and positive emotions, as well as assertiveness, energy, and excitement-seeking. High extraverts can appear likable, confident, or even brash; introverts can appear as more quiet and shy.

Agreeableness: Agreeableness has elements of politeness, cooperation, and modesty, as well as compassion, trust, and tendermindedness. Agreeableness ranges from niceness and politeness on the positive side to argumentativeness and self-centeredness on the low side.

Neuroticism: Neuroticism has elements of volatility, such as reactive anger and impulsiveness, and of vulnerability, such as sadness and self-consciousness. Neuroticism is sometimes referred to at its opposite pole as emotional stability.

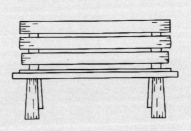

CHAPTER Four

OCEAN IS EVERYWHERE

William stood in the coffee shop, waiting to collect his coffee for a second meeting with Professor OCEAN. His thoughts were bouncing back and forth. The personality scavenger hunt had been interesting, and he couldn't wait to share the results with Professor OCEAN. Eager to get their meeting started, William grabbed the two coffees and crossed the street to the quad. Though he was ten minutes early, he rounded the corner to find Professor OCEAN seated with Sigmund by his side. Professor OCEAN looked up to greet him.

"William! How did your scavenger hunt go?"

"Well," William began, "it wasn't much of a hunt!"

"What do you mean?" Professor OCEAN asked with a knowing smile.

"OCEAN is everywhere!" William exclaimed. "I mean, it's like you gave me the red pill from *The Matrix* or something. As soon as I started looking, personalities flew at me like scrap metal to a magnet!"

Professor OCEAN didn't seem surprised at all by this revelation.

"I stopped back by the coffee shop for more fuel for the ride home last week and stood in line behind the lead singer dude from that local band that just hit the top of the charts. Of course, everyone knew who he was, so they were asking him questions. He seemed more excited to talk about the new music he was writing than the stuff that's on the top of the charts. Plus, he had lots of cool tattoos. So, I'm guessing he has a high openness score, right?"

Professor OCEAN chuckled, pleased with William's observation. "You nailed it. Perfect! Go on!"

"Before that, right outside the coffee shop, there was this lady—flip flops, shirt tail untucked, a bit disheveled. She was leaning against the wall smoking a cigarette. Everyone knew her, too, and they actually called her

Professor Cortex. They were giving her a hard time about having low conscientiousness. I thought I was in the twilight zone! It was almost like you orchestrated the whole scene! What are the odds of another personality professor showing up?"

Professor OCEAN explained that the odds, in fact, were pretty good. The coffee shop was across the street from one of the world's top psychology departments—the same department where both he and Professor Cortex worked.

"She's a genius and tremendously successful, but definitely low on the conscientiousness scale," Professor OCEAN confirmed.

"I couldn't believe it! She was making fun of herself for low C," William said.

"Fun seems to follow her!" Professor OCEAN said. "And what about E?"

"That one was easy too," William said. "I simply looked in the mirror. I'm about as outgoing as they come!"

"How about A?" the Professor prodded.

"That one was easy as well," William replied. "Mariah gets along with everyone and is really cooperative. But still, she and I butt heads because I tend to be a little lower on agreeableness."

Nodding his head in approval, Professor OCEAN pressed on. "And did anybody stand out as low in N?"

"The barista at the coffee shop," William replied. "That guy is amazing. Music blaring, people everywhere yelling out orders, and the dude was just cool as a cucumber. He kept making the coffee with no emotion. And he put an outline of the school mascot in every foamy drink! I'd have been pulling my hair out!"

Professor OCEAN smiled with approval. "Great work, William. That's the fastest you've ever done any homework!"

William could barely contain himself. "Well, it gets even better. OCEAN is all I could see all week. In every conversation, in every sales call—it's like personality was screaming at me."

"I like to call it a different lens, William," Professor OCEAN explained. "I simply gave you an updated lens through which to see the world. Now that you know it, you'll see it everywhere. Case in point: My former student,

Sophia, worked for a big tech company. She was super smart and knew psychology as well as programming. She told me how personality leaves traces throughout our social media world. You can see it in the words people use. Low agreeableness people like you, William, swear a lot more, whereas agreeable people like James use words like "grateful" and "friend" more often. You can even see it in our music choices. Argumentative people tend to listen to hard rock and rap, while agreeable people listen to artists like Taylor Swift and Wayne Newton. OCEAN is everywhere."

William paused to consider this. When he thought about it, personality really did seem to be everywhere—in all aspects of life.

"I always thought if someone wasn't like me, something was wrong with them," William told the Professor. "But this week, I simply saw people for what they were—no judgment. Everyone was different, but I could see how they all fit together, like plants in a forest. I even started thinking what the world would be like if everyone was like me. Sure, it would be better in some ways, but I can also imagine a world with a lot more conflict and speeding tickets, and no families!"

William paused for a moment. After only one meeting with Professor OCEAN, the world looked completely different. He understood himself and others in a new way. With information this valuable, he had to wonder why it wasn't something everyone knew about. Why wasn't everybody using OCEAN in their own lives?

"How come I've never heard about this stuff? Not at work or in my personal life. It's always Myers-Briggs and Enneagram, but never OCEAN."

"It's a fair question," Professor OCEAN replied. "Both Myers-Briggs and Enneagram are fine measures, and in some ways, they're easier to communicate. They put people in boxes that are simple to understand. However, the downside is that when you put people in boxes, people often won't stay in them for long. If they're close to the cut-off point, they'll retake the test and find different results. OCEAN is where the science is. That means it sticks, even if it isn't as popular."

William nodded, seeing the truth of

the statement in his own life. During their relationship, Stacey always talked about the Enneagram. He took the test a couple of times and got different results both times, each one close to the cut-off point for his designated number. It was confusing, and he never truly felt like it helped. Now, he understood why.

"Whatever tool one uses to better understand themselves is a good tool," Professor OCEAN continued. "Nothing is perfect, of course. You just need to know the limits of what you're working with. The importance is in the looking. You see, most people avoid looking under the hood of their personal make-up. Remember, you must be clear on who you are, what you want, and what your plan is to get it. There's a reason we start with discovering who you are. It's hard to know or get what you want if you don't even know who you are! You have to do the work of looking at yourself first."

William nodded knowingly. "Stacey and I used to get in big arguments over the Enneagram. I told her it was crap, and honestly, that really is what I think. It felt like she was using it to try and change me or cage me in.

And in some ways, this OCEAN stuff seems kind of similar."

William paused, taking a breath before asking the question that had been weighing on him all week. "If I do the looking, is it going to change me?"

Rather than give an answer, Professor OCEAN simply nodded. "That's a good thought to end on for today. Keep doing your own research, William. But remember, most people don't get past this first question: Who are you? Do you have the courage to keep looking?"

Then, Professor OCEAN stood, gathering his things to leave.

"You can always take a look at my website for a reading list… if you have the courage!"

And with that, he turned to go, Sigmund following along without a word.

OCEAN Is Everywhere

OCEAN Is Everywhere

OCEAN is everywhere—in music, on social media, at the office. It's not magic, but it does provide clues to who we are. It gives us a good way to talk about personality, helping us predict how people around us might think and behave.

During the scavenger hunt, William found clear examples of each of the OCEAN personality traits. He also learned that he could see those traits in almost everyone he interacted with. Can you identify the OCEAN traits of the people closest to you?

To help you begin, take a few minutes and complete the short OCEAN personality questionnaire at Professor-OCEAN.com.

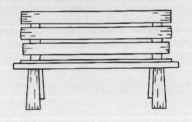

CHAPTER
Five

OCEAN IS SCIENTIFIC

During the week, William decided to dig in on the research to see if this personality stuff was really worth it. He checked out the old man's website, surprised by its simple sophistication. He used Google Scholar (a platform that he had no idea even existed the day before) to browse through all the research. He had to admit: It was impressive. He even saw a few articles from the Professor. Maybe this was really credible stuff after all.

The more he studied, the more he realized that OCEAN traits can be linked to all sorts of things. William was stunned by a paper that mapped out the personality of the United States, indicating how each state had its own predominant personality. He read on to discover more about each trait. Extraverts have more friends in real life and more social media followers. Open people are drawn to mountain areas and plant medicine. Neurotic people have trouble in relationships and a tendency for depression. Conscientious people are healthier and even live longer. More agreeable people are more likely to read religious books and share photos on social media. In short, the information was amazing!

So, William decided to give the OCEAN profile on the old man's website a try. As he answered the questions, he thought they seemed too simple and too few to accurately measure anything about a complex personality. However, as he stared at the results, he found himself agreeing that the profile did, in fact, sound like him: moderate O, moderately high C, high E, moderately low A, moderate N. At a minimum, OCEAN pegged his outgoingness and argumentativeness. It really was amazing that a short questionnaire could give him such good insight into who he was. Then again, this simple quiz was covered by more than 100 years of research to back it up. Maybe it just seemed easy because it didn't involve an expensive machine or people with white coats.

With a bit of a knot in his stomach, doubts and questions began to flood into his mind.

How can you tell anything by simply answering a few questions? If this is so great, how come Stacey has never heard of it? Why does my office choose Myers-Briggs instead? That has to be better, right? Would I get different results if I took it again tomorrow? Probably. But what if this is somewhat accurate? What if, at least at my worst, I am sometimes a narcissistic idiot?

Can personality change? Or will the worst parts of me always be this way?

Trying to shake it off, William decided to let it go for now. At least he looked under the hood a bit this week. That took courage, and he would take the rest to the old man at their next meeting.

Later that week while standing in line at the coffee shop, William wondered what the old man's OCEAN profile was like. He seemed to have a fair level of openness. His appearance—perfectly trimmed beard and fit as a fiddle—screamed conscientiousness. He seemed outgoing, but his insistence on meeting on an

OCEAN Is Scientific

obscure park bench and the fact that he lived alone made William wonder about that. He was nice enough, but William remembered the steel in his eyes when he answered questions during class. That, paired with the way he would sometimes look right through you when you said something he didn't agree with, made William wonder about his agreeableness. While he seemed emotionally stable now, William knew the fact that he was old probably contributed to that. Maybe Professor OCEAN had a little neuroticism somewhere in there.

The barista put two giant black coffees with extra shots of espresso on the counter, breaking William out of his thoughts. He grabbed them and headed to the quad. Like clockwork, there sat Professor OCEAN, legs crossed, arms on the back of the bench, and Sigmund sitting by his side smiling. Wasting no time, William went straight to work explaining what he'd done that week.

"Well, I probably did more personality reading this week than I did the entire semester in undergrad," he told the Professor. "I even took the personality profile thingy on your website."

At that, Professor OCEAN perked up. "Nice work! What did you learn?"

"Well, I guess I'm buying that personality might be a real thing that can be measured."

"Good! What else?" Professor OCEAN asked, keeping the momentum going.

"My agreeableness score was kind of low, which I guess explains folks telling me all my life that I'd argue with a brick wall."

Professor OCEAN chuckled. "I imagine the lower A trait helps you hang in there with the clients on sales calls. So, it's not all bad."

William paused, realizing he hadn't considered the positives of this (or really any of his traits!).

"Remember, we touched on this in one of our first meetings. Each of the OCEAN traits—whether high or low—can be a positive or a negative," Professor OCEAN reminded him. "It's the extreme high or low scores that we need to be aware of so that we can work with them."

"Work with them?" William asked.

"Look, we're all wired a certain way. If there

were only one personality setting that worked, we would all have it. There's a range because a trait can be really useful sometimes and not useful other times. I'm very open and curious. That looks great now, because I'm a professor who's not so worried about paying the bills. But when I was younger, my openness would sometimes take me off course in my work. I wasn't as productive as some of the more conventional people I knew. So, while I love my openness now, it has certainly been a tradeoff along the way."

Professor OCEAN paused to let William consider this for a moment before continuing.

"I guess it's like that old expression about poker: You've got to play the hand you're dealt. You have to look at your environment, see what you're up against, then maybe swap a few cards. Your wiring is important, but it's also something that you can work with, either by changing or by putting yourself in a situation where your personality really pays off."

William liked this thought. For the first time, this personality stuff didn't feel like it was meant to box him in.

"Let's take Roy, the coffee barista, for example," Professor OCEAN continued. "Roy actually owns a whole chain of coffee shops. Years ago, he wanted to expand and add additional shops. While his coffee is great and his system for getting the beans roasted and delivered is an engineering feat, Roy is not an emotional guy. He stands there like a stone pillar in the middle of all the chaos. That doesn't generate a lot of excitement. But once Roy understood that about himself, he hired someone who's wired differently to run his marketing department. The rest is history! Once Roy understood the hand he'd been dealt with his unique personality traits, he was able to use that knowledge to build a better life."

William laughed, thinking Professor OCEAN now sounded like he was selling the power of positive thinking. What really was the difference between that and all this personality stuff? Wondering aloud, William asked OCEAN that very question.

"The inside of a human is a wondrous and mysterious thing, William," Professor OCEAN answered, "but sometimes you just want to boil it down into something you can understand. It's the same as when economists look at the massive economy. They boil it down to a few simple variables to make sense out of it all. We don't mistake the map for the terrain. Roy's coffee shop is a real place, not a fraction of a number in an economist's table. And Roy is a real person, not five variables on a personality test. But we use the map to see the big picture all at once and to plan where we want to go."

"So, OCEAN is the map?" William asked.

"Yes, OCEAN is the map," the Professor confirmed. "Your personality as a whole is the terrain. You are way more than just a trait or a combination of traits. Your life is complex, filled with ideas, memories, relationships, thoughts, fantasies, and visions. To begin the journey of understanding ourselves, we have to start somewhere. OCEAN gives us a beginning point. It's sort of a 'You are here' on the map."

William paused to consider this thought. "But if it keeps changing, how is measuring personality helpful?"

"That's what is so interesting about OCEAN! Your traits are relatively stable. If you ever settle down and have kids, you'll learn this firsthand. Each one of the little curtain climbers comes with a different set of traits that you'll have to figure out. Your only other option is to go crazy trying to turn them into you. My guess is you've been argumentative from a young age, huh?"

Before William could respond, the old man hopped up and was gone, Sigmund trailing a few steps behind. As he sat there reeling from both the conversation and the abrupt exit, William wondered if personality could change. He was certain it could be measured. And now, as Professor OCEAN explained, he knew it was something he could understand. He'd discovered the hand he'd been dealt—that was the first step. But could he work with it? Maybe even change it?

OCEAN Is Scientific

We can measure OCEAN personality in a scientific way. That means you can know the hand you are dealt and play to win! William finally dug in on his personality and Professor OCEAN's personality quiz, and he learned a thing or two about himself in the process. He also had some ideas about what he would like to change.

Did you take Professor OCEAN's quiz? How did your OCEAN results motivate you to make any improvements?

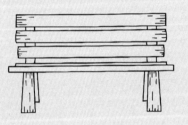

CHAPTER *Six*

OCEAN PROFILES

"Good morning William!" Professor OCEAN greeted him. "You're right on time. I'm just finishing up my first coffee."

"You got your own coffee this morning?" William asked.

"Oh William, you don't think you're my first appointment of the day, do you?" Professor OCEAN teased.

Sheepishly, William realized he had imagined that he was the old man's only student at the time. Was it that Professor OCEAN made him feel he was the only person in the world that mattered? He definitely had that effect on people! Or was it his narcissism kicking in? William chuckled to himself at that thought. It showed how much he was internalizing and buying in to this personality stuff.

"I guess you kind of made me feel like I was your only client," William admitted.

"Good," Professor OCEAN replied, "that means I'm doing my job. So, how can I help you today?"

"Well, I guess I'm bought in on this personality stuff being real," William explained. "The OCEAN model seems solid—much more researched and scientifically based than some of the other stuff that's been thrown at me over the years. What I found interesting as I dug deeper this week was the emerging research on how the traits work together."

"Well, it's pretty simple, really," Professor OCEAN began. "As best we can tell, each person has a measurable amount of each OCEAN trait. Often times, people will have a trait that is more prevalent or dominant."

Almost on cue, a very tall kid walked by the duo. He was decked out in the university's latest football gear and offered a smile to Professor OCEAN as he passed.

"Let's take him for example," Professor OCEAN said. "There's a lot more to him than being tall. But my guess is that his tallness will dominate most conversations for the rest of his life. Most people see that he's tall, but few people will ever know he got a perfect score on his ACT."

William stopped, looking back at the tall stranger who'd just walked by. Did the Professor know this guy?

"Constantine is really tall, which you could call a dominant trait," the Professor continued, confirming William's suspicions. "But he's also really smart. That's just not as obvious as his tallness."

"Okay, I get the tall and smart thing," Williams said, nodding, "but how do personality traits work together?"

"Well, first, we always need to remember that each person is different," Professor OCEAN explained. "Let's take you for example. You took the OCEAN profile. What did it say?"

William thought for a moment, considering his own results. "If I am remembering correctly, my openness was moderate, my conscientiousness was moderately high, my extraversion was high, my agreeableness was moderately low, and my neuroticism was moderate."

Professor OCEAN nodded knowingly, almost as if he knew William's profile would read as much.

"It makes sense to me," he told William. "Lots of great salesmen are wired that way."

"So, you knew my scores already?" William asked, slightly surprised.

Professor OCEAN smiled, tossing his first coffee in the trash can and starting in on his second. "Not exactly, but close enough to guess. I am a professional if you didn't remember."

"So, what does my profile have to say about me being good at sales?" William inquired.

"Well, your high extraversion helps you meet people, and your low agreeableness helps you hang in there so you can turn a no into a yes. Your mid-level emotionality—or neuroticism—means you can run a little hot at times."

Professor OCEAN paused to sip his coffee before continuing.

"Your issue is that you are wired to be a short-term seller more than a long-term relationship builder. A little more agreeableness might make for longer term clients and referrals. Everything is a trade-off. What makes you great in sales might make you weak in personnel management. And if you can't manage, you won't rise any higher in the organization. It really depends what life you want."

William considered this for a moment. "Yeah, as Stacey and Mariah constantly remind me, selling is not my problem. It's the playing well with others that gets me in trouble."

"Am I hearing a hint of some ownership of the problem?" Professor OCEAN teased.

William shook his head. He wasn't yet ready to fully admit the people in his life might've been right about him after all.

"So, what about James' personality?" William asked, shifting the focus.

"Well, now that you know how to look at personality through the lens of OCEAN, what do you think?" Professor OCEAN pushed back.

"Well, I love the guy," William began, "but when we were roommates in college, he seemed to be a bit of a do-gooder. And his insistence on cleaning up the apartment drove me crazy! So, I guess his love affair with keeping things in order would say he is very conscientious."

"That's right!" Professor OCEAN confirmed

proudly. "High Cs are disciplined and like things a certain way. Really high conscientiousness can even become a bit obsessive compulsive."

"Oh my gosh! Thank you for saying that!" William exclaimed. "Living in all that neatness at times seemed like hell to me!"

"I didn't say James was OCD," Professor OCEAN cautioned. "You have to remember that James has his own road to travel and that road is different than yours."

William paused to consider this for a moment. All he heard in college was that he should be more like James. But to William, James never seemed to have any fun. Well, other than football games on Saturdays, of course.

"I guess in hindsight, he was even disciplined about his fun," William told Professor OCEAN. "That's why it was such a surprise when he showed up with a tattoo a couple of years after college!"

"Why was that a surprise?" Professor OCEAN asked.

"Well, he just seemed too disciplined to do something like that," William explained. "It seemed out of character."

"Maybe you just didn't know James as well as you thought you did," Professor OCEAN suggested. "New ideas and experiences are rooted in the O—openness. My guess is that James is a little more open than you give him credit for."

William considered this for a moment. Maybe Professor OCEAN was right. James quitting his job a few years ago to start his own company felt like another example of what Professor OCEAN was saying. This personality stuff really was helping him understand the people in his life in a new way.

"So, what about Mariah?" William asked. "She seems different than James, but she tells me the same things James used to say when we were roommates."

"Well, since you're learning so much, let me ask you again: What do you think?" Professor OCEAN replied.

"Well, she seems really nice and she is a hard worker," William considered. "However, I seem to trigger her with some of my stuff. When I challenge people in meetings, it seems to make her emotional. She doesn't say anything, but I can see her face turn red."

"What you are describing could be the combination of high agreeableness and moderately high neuroticism," Professor OCEAN explained. "With this combination, the high agreeableness seems to rule the roost, but there is also high emotionality."

"I've never understood how Mariah made it in sales," William confessed. "I've sold more in one year than she sold in her entire career. And now she manages me!"

"Am I hearing some resentment?" Professor OCEAN pushed back lightly.

"Yes," William admitted. "Who is she to tell me what to do? She stunk at sales!"

Professor OCEAN leaned in. "William, have you ever considered that maybe she has the perfect personality for her job?"

William stayed silent in response.

"It seems to me that her job is not to sell; her job is to prevent her top salesman from burning down the joint." Professor OCEAN explained. "That sounds like a perfect job for Mariah's personality. A little water for the fire, maybe?"

William was speechless. Once again,

Professor OCEAN had given him an entirely new perspective to consider.

Taking a final sip of his coffee, Professor OCEAN took William's silence into account and said, "Well, it seems like my work here is done for today!"

William stopped him short. "Now wait a minute, before you pull that disappearing act of yours, I have a few more questions. What about Stacey? What does her OCEAN profile look like?"

"Why do you care?" Professor OCEAN asked knowingly.

With his motives exposed, William responded quickly. "I didn't say I cared! I'm just wondering."

Professor OCEAN paused, raising his eyebrows before putting the question back on William once again.

"What do you think?" the Professor asked.

"She seems kind of like James but not quite as neat and without the tattoo. And she's more fun."

William paused to consider what this might mean.

"Maybe she's a little lower on O than James, a little lower on C, and a little higher on E?" he offered.

"Makes sense to me!" Professor OCEAN exclaimed. "That's the great thing about OCEAN. You don't have to perfectly categorize someone to understand them. We can understand people simply by comparing and contrasting them with others. Again, you just proved my work is done here, so I'll see you next week!"

And with that, the old man hopped up and threw the coffee cup in the trash can from long range. (His hand-eye coordination was still quite impressive!) In a flash, he was off with Sigmund in tow.

OCEAN Profiles

OCEAN Profiles

The OCEAN lens is a reminder that you are not one trait. You are a combination of all five traits, which work together to form your personality. William's mix of high extraversion and low agreeableness helps him to be a great salesman, even though he's not always the nicest person. Without his high level of extraversion and enthusiasm, William would probably be a guy you'd want to avoid!

So, consider this question: How do your unique "Big Five" traits work together?

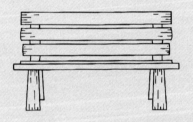

CHAPTER Seven

PERSONALITY CAN CHANGE

After noticing that William had been a little more agreeable in the team meeting that week, Mariah stopped by to check in on him at the office.

"How's it going?" she asked.

"I'm on fire this week!" William exclaimed. "I just closed that deal I've been working on for the last six months."

Mariah was surprised. "I thought the client rejected your proposal a few months ago."

Smiling, William continued. "She did! But with the new price change, I decided to give it one more try."

Mariah laughed, realizing that sometimes William's hard-headedness actually paid off.

Personality Can Change

Maybe that time with Professor OCEAN was changing the way he let his personality work for him after all.

"So, how's it going with the Professor?" Mariah asked. "Did you cover the way personality can change last week?"

She was expecting a "yes" based on what she'd observed from William's performance in the team meeting.

"Nope, we never talked about that," he answered. "We got to talking about the way OCEAN traits work in combination. It was so interesting that we never moved on."

Mariah smiled knowingly. William wasn't the only person who bought coffee and sat on the park bench with Professor OCEAN. She knew those lessons all too well herself.

William considered this as he drove to his weekly meeting with the Professor. James knew Professor OCEAN's work well. It seemed Mariah did too. Who else in his life would he discover had unlocked the science of OCEAN?

For his part, William now saw how the OCEAN profiles made perfect sense. He had always thought he had the "right" personality, meaning everyone else's was "wrong." Now, he was seeing it all differently. For the first time, he saw Mariah in her role as manager differently than he had in the past. OCEAN was changing the experience at work and his relationship with his boss.

Walking up to the park bench with coffees in hand, there sat the old man—no coffee, arms spread across the back of the bench and legs crossed while talking to Sigmund, who seemed to be listening attentively. As William approached, Professor OCEAN perked up.

"Good morning William! What's new this week?"

"I had a great week!" William explained, settling in beside the Professor.

"Really?" Professor OCEAN asked, almost as if he already knew. "Tell me all about it."

"Well, it was really interesting. After our conversation last week and a little more research, I seemed to be looking at situations from other points of view, not just my own."

Professor OCEAN nodded approvingly, encouraging William to continue.

"In the weekly team meeting, I used to sit

there thinking about what idiots the other sales folks are. This week, I caught myself thinking about each person's OCEAN traits. Simply looking at them through the OCEAN lens helped me consider how their points actually made some sense. One of the salespeople was going on and on about mindfulness meditation. I used to think he was kind of a weird loser, but when I applied the OCEAN lens, things made more sense. He is clearly conscientious, disciplined, and hardworking. He is a little more agreeable than I am, but he is also kind of neurotic. He's just a little bit socially anxious whereas I'm socially confident. He was talking about how every other salesperson could benefit from mindfulness. I guess he assumed they were anxious and disciplined like him, just like I assumed every salesperson was fearless and a bit arrogant like me. It seems the world is a little more complicated than I thought."

"Impressive observations, William!"

Professor OCEAN was obviously pleased with his student.

"So, what you're saying is maybe there isn't just one right personality?" the Professor pressed.

"Exactly!" William confirmed. "Your comment last week about Mariah's job being to keep the extraverts from burning the place down hit a nerve. She certainly has saved me from myself lots of times! I see it all so differently now."

Professor OCEAN took a big sip of his coffee, glowing at the progress he was seeing in his student.

"So, the big question: Did my personality change this week?" William asked. "Is it possible to change personality?"

"It sounds like you are ready for our final lesson," Professor OCEAN said.

William was surprised to hear their meetings might be coming to an end. He was even more surprised at the twinge of sadness he felt at the thought of it.

"The answer is yes! Your personality did change this week," Professor OCEAN confirmed. "And yes, personality can change. But I say that with a caveat."

William was intrigued. "Meaning?"

"There are a couple of ways to think about personality change. Most of the time, personality change takes some effort or help. The most obvious example of this is talk therapy or psychotherapy. It used to be that anyone seeking therapy had some type of major issue. So much so that admitting you'd had psychotherapy was a badge of shame. Now people hire a coach or go into treatment because they're working to simply improve their personality, much like fitness training."

Here, the Professor leaned in before continuing. "Perhaps a person wants to be less neurotic. You know, William, the antidepressant medicines that many people take actually work to lower neuroticism. The same pill works on different disorders like depression, anxiety, and OCD because those disorders are all rooted in neuroticism. Or maybe a person wants to be more agreeable—more loving with their children or spouse. Therapy can help. There's lots of research showing that talk therapy is an effective way to change personality. It just takes time."

William nodded, considering Professor OCEAN's words.

"Other people will change their personality

through work or practice," the Professor continued. "Your friend at work who is practicing mindfulness is actually making himself a less neurotic person. It's also probably making him a better partner in his marriage. It just takes effort and work."

Professor OCEAN paused to sip on his coffee, letting William mull this information for himself.

"Another way to change yourself is by changing your social group," the Professor explained. "William, I'm guessing you're not hanging out with the church choir. Maybe if you spent some time around people who are more loving or engaged in service to others, your own agreeableness would expand. This is true for all the traits. If you want to become more open, find yourself a group of creative, interesting, or adventurous people to spend some time with. It might be a little unsettling at first, but given time, you might find yourself becoming much more creative and interesting because of it."

William leaned in, interested at the potential of change as the Professor went on.

"Finally, there is another kind of personality

Personality Can Change

change. This one is rare and not always pleasant. They call this quantum personality change because it's so rapid. Unfortunately, one place this occurs is in warfare, where trauma can alter people's personality in an instant. The experience makes them more socially anxious, fearful, or neurotic. The good news is another kind of quantum change—a mystical religious experience sometimes aided by ancient plant medicines—can lower the neuroticism again. Of course, quantum personality change is really interesting, but for most people like you, William, change is going to be more about practice."

"So, let me get this straight. Personality can change," William repeated, "but for most of us, it happens slowly. Did learning to see the world through other points of view change my personality?"

"Theoretically, yes!" Professor OCEAN confirmed.

"That seems too simple, Professor," William replied.

"Simple yes, but easy, no." Professor OCEAN distinguished. "That's why so few people do it!"

"So, things like meditation, spiritual practices, talk therapy, pharmaceuticals, plant medicines—it all works?" William asked. "It all has the potential to change us?"

"Of course, they all work. Each tool can be beneficial. Each tool simply works on different parts of personality and for different people at different times in their lives," Professor OCEAN explained.

"That's helpful," William told his teacher. "It's the first time I've heard it explained that way."

"I've been explaining it that way since undergrad, William!" Professor OCEAN laughed. "Maybe you just skipped those classes."

William laughed sheepishly. "Okay, so why did there seem to be instant change in my personality last week? I mean, I usually sit in team meetings aggravated at what others are saying, but this week, not so much."

"We can't really know for sure, William," Professor OCEAN replied. "The brain is one of the last great mysteries. But my hunch is that you had a sudden flash of insight that prompted a change in your personality. This is something

Personality Can Change

we see in creativity research when people solve a puzzle. Before, there were several areas in your life where the pieces weren't fitting together. You're great at striking up new relationships but not good at having close relationships. You're great at your job as a salesman but not great as a teammate at work. When you looked at yourself through the lens of OCEAN, you saw all of the reasons behind that for the first time. It helped the puzzle pieces fit!"

William knew the Professor was right. For the first time this week, he saw that his low agreeableness and high extraversion put him in this position. He also saw that if he were a little more tolerant and considered other people's perspective, his life would still have all the upsides but without so many downsides. In a way, William felt like after just one week of self-study he'd found the insight that would take a lot of therapy to get to.

"William, I think you are in the right place. You're ripe for change," the Professor encouraged.

William sat pondering what the old man was saying.

"For the first time, you saw each of your teammates as individuals—as unique bundles of OCEAN traits and experiences with unique stories. You saw the world from their points of view. But you have to remember that even when personality change seems sudden or dramatic, often there's a lot that led up to the change. In your case, you'd been struggling with these issues of connection at work and how you could be more effective for quite a while. Your look into those struggles bore fruit quickly, but you had to spend a lot of time in the dirt to get there. That's often the way it is. You can spend months wrestling with a set of ideas and then one day, boom! It all clicks!"

"That makes sense," William said, nodding. "I think you're right."

"And I have a hunch that a new thought crept into your head this week," Professor OCEAN pushed on. "Instead of believing you have the one right personality, you realized there are other personalities that are needed to make the world go around."

William nodded, quietly confirming the lesson Professor OCEAN had taught him in

Personality Can Change

their time together.

"Regardless of how it's happening, you are doing the hard work of exploring yourself," Professor OCEAN encouraged. "For that, you should be proud, William."

William stopped at the old man's use of the word "explore." That was the exact word Stacey had used once in one of their conversations about William's personality! She'd encouraged him to "explore" it. Did Stacey know the Professor too?

"Why did you just use the word 'explore'?" William asked, intrigued and confused at the same time.

"Oh, who knows?" the Professor replied. "But it does remind me of one of my favorite quotes: 'Make a careful exploration of who you are, and the work you have been given and then sink yourself into that. Don't be impressed with yourself. Don't compare yourself with others. Each of us must take responsibility for doing the creative best we can with our own life.'"

"I love that quote. Is that a Professor OCEAN original?"

"Nope, it's 2,000 years old. Saul of Tarsus

said it."

"So, it's that simple?" William asked, moving on. "I just need to become more aware of my personality and the way it impacts others?"

"Yes!" Professor OCEAN confirmed. "Just don't forget the hard work of changing. We'll see how well your new agreeableness sticks when you are stuck in traffic after a sleepless night."

William stared off in the distance, lost in thought. He thought about his friendship with James, his relationship with Mariah, even the potential of a relationship with Stacey again. His friends, his family, his coworkers—all the relationships in his life could change for the better if he saw them through the lens of personality.

"Let me put a little more philosophical spin on it," Professor OCEAN continued, breaking the silence. "I've been thinking about these issues for a long, long time. It seems to me that the only way for us to get along as humans is to all be different, work together, and come together in interesting and important ways. Because of that, I've started to have a lot of gratitude, not

Personality Can Change

just for my obvious strengths but also for my weaknesses."

Here, Professor OCEAN paused to let his words sink in.

"To sum it all up, William," he began, "The most important lens you can acquire for seeing yourself and others is gratitude. Gratitude for your unique OCEAN traits and gratitude for your unique journey in life."

With that, Professor OCEAN delivered his closing statement. "William, my work here is done."

Professor OCEAN stood up, took his last swig of coffee, and flicked it (nonchalantly as always) right into the middle of the trash can. William stood up, almost as if he wanted to block the Professor from leaving. He didn't want their time to be over.

"Before you do your disappearing act," William cut in, "I have one more question. What's your OCEAN profile?"

Professor OCEAN turned to look at William, his blue eyes staring intently at his student.

"William, if I've done my job, you should know exactly what each of my OCEAN traits are," Professor OCEAN said. "But at the same time, you shouldn't have a clue…"

And with that, he began to walk away, Sigmund by his side. Without even a glance back, he shouted one last parting thought.

"Former students are always welcome to attend one of my lectures, William."

William stood mesmerized. Just like that, the Professor was gone.

The lesson had ended.

Personality Can Change

Good news! Personality can change! William had trouble being completely honest about the type of person he was, in part because he thought he was stuck. Professor OCEAN showed him that with awareness and effort, his personality could change. He could move his life in a better direction.

When it comes to your OCEAN profile, which ways of changing your personality are most appealing?

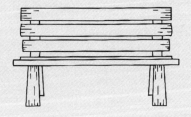

CHAPTER Eight

PERSONALITY GROWTH

The following week, William found himself back on campus. Mariah had given their team the day off leading into a holiday weekend, so he had some extra time on his hands—time he decided to spend in the last place he ever expected to be again: Professor OCEAN's class. Normally, William would have taken a day like this to play golf with some buddies or have some fun. But, missing the routine of meeting with the old man, William checked Professor OCEAN's lecture schedule and saw that Thursday was actually the first class of the new semester. The Professor was scheduled to teach a class called "Wealth Science™" at the exact time they had been meeting, so William decided to make the drive and attend.

He stopped by the coffee shop for a large

Personality Growth

cup with a shot of espresso. The coffee shop was hopping, and it was good to see Roy calmly working his magic. William walked through the quad, down the hill, and approached the auditorium with the massive football stadium in the distance. As he entered the auditorium (yes, five minutes late), Professor OCEAN was already lecturing with Sigmund sitting alertly beside the lectern. Sigmund glanced up at William as he entered, wagging his tail as if welcoming him to class.

As he moved to find a seat, William noticed more than a few familiar faces in the room. Much to his surprise, there sat Mariah and James, both intensely taking notes on their computers. He chuckled to himself. Of course they were here! As he politely entered the only row with a few open seats, he glanced to his left and immediately locked eyes with an unexpected face—Stacey. This felt like the moment of truth. It was time to see if William's personality change would result in some growth in his life after all!

The answer would come just two years later, when William and James happened to be vacationing in the same Florida beach town.

As old friends do, the two decided to meet for coffee. William walked in five minutes late to find James with what appeared to be a few extra tattoos on his left arm.

"Howdy, stranger!" William greeted his friend as he walked toward the table.

"Great to see you!" James exclaimed. "I don't think we've seen each other since your wedding last year. How's Stacey?"

"She's great," William said of his wife. "It's been a wild couple of years. And it just keeps getting wilder!"

"Your wedding was an absolute blast," James told him. "I wish we had more time together, but our company continues to grow like gangbusters. I had to get in and out quickly to keep up with work."

Laughing, William replied, "I'm just glad you could make it. I wouldn't have wanted you to miss Professor OCEAN's dance moves at the reception! Or Sigmund's doggy tuxedo!"

The two men chuckled, remembering the scene.

"The changes you've made are amazing," James told his friend. "It's really quite impressive."

William nodded, pleased with his own work and progress.

"But don't get too full of yourself," James teased. "I see you were still five minutes late today."

"And let me guess, you were 15 minutes early!" William responded.

James smiled sheepishly in return. "Some things never change!"

"But some things do," William said. "You know, the reason I wanted to meet today is to tell you the good news in person. Stacey and I are expecting a baby!"

James was overjoyed. "Man! That's amazing! Congratulations."

Here, William turned serious.

"I want to thank you, James, sincerely. You pushing me back to Professor OCEAN changed my life. It changed me for the better. My personality, my relationships, my experiences—I see and appreciate them all in a new way."

James smiled, sincerely happy for his dear friend and the progress he'd made in his life.

"I'm so glad, William. But if you're looking

for a new experience, just wait until the baby comes!"

With that, they both had a laugh, celebrating the past and contemplating the future with another new personality in the mix.

Personality Growth

Embrace growth! You can change your personality to match the life you want.

At first, William thought that growing as a person would mean losing something. But it turned out that growth gave him a much richer world. Instead of losing his edge as a salesman, he found himself enjoying deeper relationships with his customers and seeing better results. Instead of feeling boxed in, giving a little allowed him to build a life with Stacey that worked for both of them.

So, that leaves us with you.

Do you have the courage to look under the hood in your own life?

Are any of your OCEAN traits a surprise to you?

Is there a trait you now realize is a strength?

Is there a trait you now realize is a weakness?

How will your awareness of these traits change the way you see and value others?

How will your awareness of these traits change the way you approach or avoid certain circumstances?

How will your awareness of these traits change or enhance the story you tell yourself about you and your life?

You have the tools. The rest is up to you!